It's About Time!™

All About a Day

Joanne Randolph

KiDS
press™

New York

Published in 2008 by The Rosen Publishing Group, Inc.
29 East 21st Street, New York, NY 10010

First Edition

Book Design: Kate Laczynski
Photo Researcher: Nicole Pristash

Photo Credits: Cover, p. 1 © Larry Dale Gordon/Getty Images; pp. 4, 6, 8, 10, 12, 14, 16, 18, 20, 22, 24 © Shutterstock.com.

Library of Congress Cataloging-in-Publication Data

Randolph, Joanne.
   All about a day / Joanne Randolph. — 1st ed.
       p. cm. — (It's about time!)
   Includes bibliographical references and index.
   ISBN-13: 978-1-4042-3767-4 (lib. bdg.)
   ISBN-10: 1-4042-3767-4 (lib. bdg.)
   1. Day—Juvenile literature. I. Title.
   QB209.5.R359 2008
   525'.35—dc22
                                                        2006037188

Manufactured in the United States of America

Contents

A day is made up of 24 hours, with 365 days in a year.

Days are part of passing time. We'll learn about them here!

Each day starts when the morning sun peeks in.

We get up, **brush** our **teeth**, and let the day begin.

Next we eat breakfast
to give us power for our day.

We're ready to go to school
or spend the day at play.

Morning ends at twelve o'clock.
It's time to eat our lunch.

Peanut butter **sandwiches**
are really good to munch.

In the afternoon the Sun heads for the west.

Class ends, we play a sport, or study for a test.

Daylight is going, and darkness starts to fall.

We eat our dinner and maybe play some ball.

Nighttime starts when the Sun has gone away.

We do quiet things as we end our day.

At night we get ready to go to sleep.

We'll read stories or count some **sheep**.

Even though our eyes close, the day is still not done.

Nighttime animals come out to eat. Their day has just begun.

Morning, afternoon, and night are part of each day.

Time keeps moving. So do we, at school, work, and play!

Words to Know

brush

sandwiches

sheep

teeth

Index

Web Sites

Due to the changing nature of Internet links, PowerKids Press has developed an online list of Web sites related to the subject of this book. This site is updated regularly. Please use this link to access the list: www.powerkidslinks.com/iat/aday/